Backpack Bear's

Bird Book

Written by Alice O. Shepard

Starfall®

Starfall Education, P.O. Box 359, Boulder, CO 80306   ISBN: 978-1-59577-087-5

There are five kinds of vertebrates. Two of these are "**_warm-blooded,_**" and three are "cold-blooded."

This book is about birds.

Vertebrates (Animal

Mammals

Birds

"Warm-Blooded"

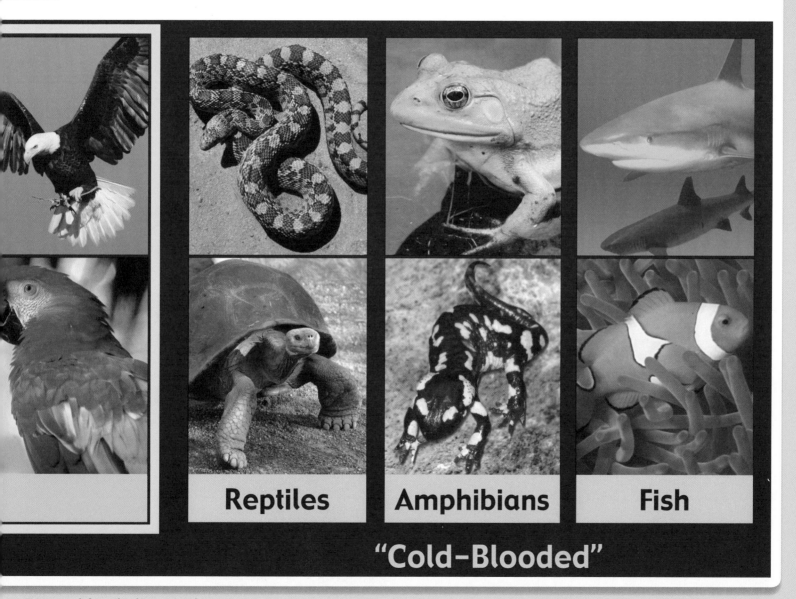

Reptiles   Amphibians   Fish

"Cold-Blooded"

"Warm-blooded" animals can maintain a constant body temperature. "Cold-blooded" animals generally are not able to do this. Their body temperature changes depending on their surroundings.

A **bird** is a type of "warm-blooded" animal with a backbone.

Birds have **feathers** to keep their bodies warm. This characteristic makes an animal a bird!

I know my pet is a bird because it has feathers.

Bald Eagle

5

Penguin

Hummingbird

Look at these two animals. Are they different from each other?

Yes, they are different from each other, but because they both have feathers, they are both birds!

Blue-Footed Booby

Parrot

Birds come in many shapes and sizes. Their feathers come in many different colors.

Birds live on every continent on Earth.

Chicken

Peacock

Owl

Frigate Bird

Flamingo

Birds have four limbs. Two of their limbs are legs and the other two are **wings**. Most birds use their wings to fly.

Some birds **migrate**. When the seasons change, these birds fly long distances, from one place to another.

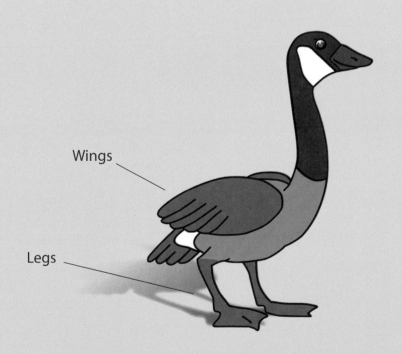

Wings

Legs

Canadian Geese

Some birds can fly only short distances.
They fly to escape predators.

Pheasant

Penguin

Ostriches

A few birds can not fly at all! Penguins use their wings to swim. They look like they are flying through the water.

The ostrich has very small wings, but long powerful legs. These birds are runners, not flyers.

13

Baby birds are called *hatchlings* or chicks. They hatch from eggs laid by the mother bird.

Most birds lay their eggs in *nests* they've built from things like twigs, string, mud, and feathers. Nests are built in high places, low places, and even underground!

In a Tree

On the Ground

Made of Rocks

On a Ledge of a Building

Blue Jays Feeding Chicks

Some bird parents find food and bring it back to the hungry hatchlings. They work all day long to keep their babies fed.

Ostriches

Ducks

Birds take care of their babies, protecting them and teaching them how to survive on their own.

Most birds eat insects, seeds, or fruit.

Some birds eat fish.

A few birds, called raptors or **birds of prey**, hunt and eat smaller animals.

I like to eat fish, too, but I am not a bird.

Bald Eagle

Seed-Crushing Beak

Mud-Probing Beak

Raptorial Beak

Insect-Catching Beak

Fruit-Eating Beak

Chiseling Beak

Nectar-Feeding Beak

Dip-Netting Beak

All birds have **beaks**.

A bird's beak gives us a clue about the kinds of food it eats.

21

Wading Feet

Webbed Feet

A bird's feet tell us where it lives and how it gets its food.

Almost all birds have four toes, but not all birds' toes point in the same direction!

Talons for Hunting

Scratching Feet

Perching Feet

Climbing Feet

Some birds use their feet to grab and hold onto things the same way humans use their hands.

Many birds live and feed together.

They stay in groups to protect themselves and their young from predators.

Flock of Red-Winged Blackbirds

Gray Catbird

Oriole

Red-Winged Blackbird

Sandhill Crane

Birds communicate with chirps, calls, hoots, quacks, squawks, and songs. Some birds sing beautiful songs.

I like the red-winged blackbird's song best of all!

Bald Eagle

Turkey

Pigeon

Vulture

All birds have wings, but not all of them fly.

All birds lay eggs and most build nests, but their eggs and nests do not look the same.

All birds have beaks, legs, and feet, but these body parts come in many shapes and sizes, and are used in many different ways.

With all these differences, how can we tell if an animal is a bird? Remember: An animal is a bird if it has feathers!

Glossary

Beak: A bird's hard jaws

Bird: A "warm-blooded" vertebrate covered with feathers

Bird of Prey: A bird with a hooked bill and talons that hunts and eats small animals

Feathers: The covering of a bird that helps keep the bird warm

Hatchling: A young bird recently hatched from an egg

Migrate: To move from one place to another as the weather changes

Nest: A structure or place made or chosen by a bird for the purpose of laying its eggs

"Warm-blooded": An animal that can keep its body temperature nearly constant even though the temperature outside may change

Wings: Two limbs of a bird, often used for flight

Can you guess what animals these are by looking at their feet?

If you guessed a T-Rex and a chicken—you are right! Look closely at their feet. Do you think the T-Rex and the chicken might be related to each other? As strange as it may sound, most scientists agree that dinosaurs and birds are actually cousins!

For many years, scientists have had the theory (idea) that dinosaurs and modern-day birds are related, but they did not have proof. In 2000, a Chinese farmer found a 125-million-year-old fossil (an impression in a rock) of a dinosaur with feathers! His discovery proved that dinosaurs are closely related to birds.

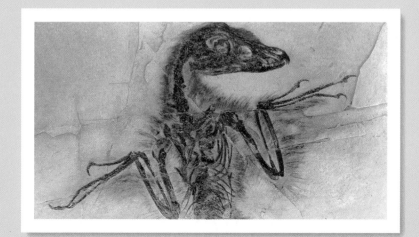

Little "dino-bird" Dromaeosaur
Hundreds of "dino bird" fossils like this one were found in northeast China. You can clearly see the fine feathers, proving that dinosaurs and birds are cousins!

Index

About the Author

Alice O. Shepard and her mother Elizabeth loved to watch birds eating from the bird feeders outside their home in Southampton, England. She and her mother played a game to see who could name the most birds. They kept of list of all the birds they saw and named. It was always a very big deal when either of them saw a rare or unusual bird. Many people in England play this game. Whoever has the longest list of birds is the winner!

Acknowledgements

Special thanks to Rebecca J. Safran, Assistant Professor, Dept. of Ecology and Evolutionary Biology, University of Colorado, and to Byron Swift, for helping to check this book for accuracy.

Photo Credits